Why MATH?
Mental Anguish To Humanity

Engaging At-Risk Students in Math & Science When the Teacher Hates Teaching It!

Why MATH? Mental Anguish to Humanity: *Engaging At-Risk Students in Math & Science When the Teacher Hates Teaching It!*

Anchor Book Press
440 W Colfax Street, Unit 1132, Palatine, IL 60078
ISBN: 9781949109085
Cover Art by Ben, 8 years old

Printed in the United States

Other Books by Belinda Adams

Can You See Me? *Using Understanding to Help Students of Poverty Feel Seen, Heard, and Understood in the Classroom*

Can You See Me? *Quick Start Implementation and Reflection*

Don't Look Too Closely: *What Students of Trauma Are Hiding & How You Can Be the Difference for Them*

Don't Look Too Closely: *Quick Start Implementation and Reflection*

Why MATH?
Mental Anguish To Humanity

Engaging At-Risk Students in Math & Science When the Teacher Hates It!

Belinda Adams

Anchor Book Press · Palatine

Dedication

This book is dedicated to my friend and publisher/editor/ marketer, Carol. I can honestly think of no one who admittedly loves math as much as this woman! She encouraged me to write this book after watching me struggle my way through for a few years, and knowing that, regardless of my lack of passion for the subject, I found ways to make the lessons understandable and engaging (for my students and myself).

Because of her encouragement, it is our hope that other teachers who have their own "teaching kryptonite" will search for ways to enhance their instruction and engage students in all subjects. Thank you, Carol!

Acknowledgements

I'd like to acknowledge my students for the moments (or hours) they have listened, watched, and learned alongside of me on academic subjects that were not my favorite. Often, through their excitement, I, too, discovered that math doesn't have to be boring nor does science have to be cumbersome.

"...Adjust the curriculum and presentation of information to learners rather than expecting students to modify themselves for the curriculum."

The Universally Designed Classroom

Forward

Belinda has done it again. She has written another book that touches the hearts and souls of readers. First, she shared with us things that have worked in her classroom in her books, *Can You See Me* and *Don't Look Too Closely.* Now, Belinda shares with us the things that have worked to help her become a better teacher of the subjects she would rather not teach.

I'm sure all primary teachers can identify with the feelings of reluctance to teach subjects they do not like. While I loved math as a kid and enjoyed teaching it, I can't say the same about science. Sure there are some fun things in science, but who really wants to teach about the intricacies of cells. Ok, some of you do, but there are other things you are not jumping for joy to teach. Belinda gives strategies to help us overcome our reluctance and prepare engaging lessons for our students.

Why Math is targeted at math and science, but the ideas can be used for any subject you are not thrilled to be teaching. What? you say you are in a professional learning team where everyone has to do the same thing. Great, your colleagues will love the suggestions you contribute to the group after reading this book.

Many times, when we see something new, we can't help but think what's in it for me. Sure, none of us went into teaching with that in mind, but sometimes when we are struggling with this student, that new curriculum, or even personal issues, it's ok to say what's in it for me. If you have ever felt that way, then this book is definitely for you, because there is something here for you, the teacher. Yes, it is all about the kids and what is good for them. However, as you read this book, you will find sometimes, what is good for the teacher is also what is best for students.

One thing I learned while teaching with Belinda was that she was always willing to do whatever it took to help her students. Whether that was searching the internet for videos on how to teach angles or checking with a teacher who loved science to see what she was doing to make gravity fun and exciting or looking at that one kid and trying to figure out just what he needed to be successful. No matter what was needed, Belinda would look for the answer. She realized that some strategies were used over and over again, only the subject or kid and his story had changed. She has taken those stories and written about them in this book to share with others who are willing to do whatever it takes.

I must say I am impressed with the effort Belinda puts into teaching subjects she doesn't like and at the number of students that she is able to engage that at one time would have just tuned out because they didn't like the subject. This book is a quick read, but it will give you strategies you can use to plan lessons this week and ideas when issues with students arise in the coming months. Overall, *Why Math? Mental Anguish to Humanity* is a compelling book that has practical strategies that will not only help your students but will help you in terms of greater student achievement and fewer struggles with students' behavior.

Glance back at the cover art, for it is a visual that sums up our goal – to prepare a lesson that prevents students from realizing you don't like the subject and inspires the student to feel good about the subject.

Carol Pirog – educator and author

Table of Contents

Introduction

I am a "big picture" kind of individual. Unlike those who enjoy the details – dotting the "i's" and crossing the "t's" of each letter in the word, I like to envision the entire word, or better yet, the entire paragraph. I'm often teased for wanting to know the end of the story first, so I can enjoy the story being told.

When it comes to doing a project, I want to know "what's the end product going to look like?" and then, let me work backwards to plan the steps which need to be completed to create the end product. Notwithstanding, I know my polar opposites who want to painstakingly "map out" each step while envisioning the end product.

Every time I answer my cell phone or utilize the navigation system in my car, I marvel at the wondrous minds of those individuals who see "details" as "gems" to be cultivated, shined, and applied to just the right masterpiece. Without their exquisite abilities to complete math problems I'd find difficult copying onto notebook paper, I would not enjoy many of the luxuries I take for granted.

I often think about how blessed all students would be if each of their math and science teachers could be mathematicians or scientists. Wow, mind blowing! right? They'd be terrific and excel in those high-paying fields.

However, those scientists and mathematicians most likely would not fare well in classrooms where little 1st graders wipe their noses on their sleeves, slurp their milk at lunch, and sneeze on the teacher's desk! I smile when I think about my friend's husband who works for Motorola and has such a brilliant mind! He'd last about one day in that environment before he called it quits, and I don't think he'd be offended to know I said that!

So, if you're like me and millions of other educators who love your profession, have a passion for children, and yet, can't seem to locate your love for all subjects, this book is for you!

It's in the Curriculum

I was fortunate when I taught 5th grade special education for 9 years. I shared the curriculum with a partner teacher. We were able to break up the curriculum and "flip" our two classes in the afternoon.

I taught Reading, Language Arts, and Writing, while my partner teacher taught Math, Social Studies, and Science. It was a wonderful, productive teaching relationship!

During this time in my teaching career, my students (and hers) made phenomenal growth, both in reading and math. Why? Because each of us was teaching what we enjoyed, and consequently, we became great at our craft!

What happens, however, when you cannot "divvy" up the curriculum? This is especially true for the primary grades. When I taught general education for the lower primary grades, I had my students all day and taught all subjects. Similarly, when I taught in a specialized emotional disorder classroom, I was responsible for all subject areas.

Fortunately for me, both of these situations involved the lower primary grades.

Why might that be important? Well, for someone who doesn't enjoy teaching math and science, it's so much easier to teach math and science in the primary grades. It would be nearly impossible to screw up teaching how to tell time, how to count money, or seasons of the year!

However, you can ask my partner teacher, it was pretty easy to mess up 5th grade math. We still laugh at the day she was helping a student correct a wrong answer, when he asked, "how can it be wrong, that's the answer Mrs. Adams told me". In my defense, this was the new Common Core math and it wasn't that I was unable to do 5th grade math, I just didn't understand the Common Core way of doing it.

Even in the lower primary grades, while I knew the material, I had to find methods and strategies and lessons that were engaging, both for myself and the students. This is integral because students have a well-tuned radar for teachers who are instructing them in curriculum they themselves do not find interesting or engaging.

As I've stated more than once, if it's important to the teacher, it's important to the students! The reverse is just as true, if students know it is unimportant to the teacher or the teacher doesn't like the subject, they won't put a lot of effort into learning.

This book is dedicated to teachers like me, who develop strategies to engage her students (and herself) in subjects that are not appealing in order to encourage them to learn and increase academic achievement. While the stories in this book are targeted to math and science because those are the subjects I do not enjoy teaching, the strategies can be used for any subject.

Learning is supported and facilitated by the interaction of the learner and the curriculum. When the support and inclination is missing, 'learning disabilities' arise.
If the curriculum cannot be flexibly designed, it cannot meet all learners where they need to be met."

The Universally Designed Classroom

Task 1: Identify Your Teaching Kryptonite

I believe it's difficult for us, as educators, to admit that we're not great at teaching all subject areas. We like to think that we are trained professionals who have been through multiple years of college and professional development which has prepared us to teach everything a child needs to learn. And parts of that statement are true.

However, if we are honest with ourselves, I'm sure each of us can identify a curriculum area that just doesn't make our creative hearts burn with fire. For me, it's math. Followed closely by science. I can generate some enthusiasm for social studies; however, I particularly enjoy language arts units that involve my creative efforts for things like the Midnight Ride of Paul Revere or The Underground Railroad. Ask me to teach about types of angles or the different rocks and minerals, and you'll find I'm not so enthused.

I'm not saying that I don't realize they are important subjects for students to learn in school. I just have to be honest with myself and admit that there are certain subjects that are not appealing to me as a teacher. I didn't enjoy them as a student and things have not changed now that I am the teacher. Consequently, I am sure that my lack of enthusiasm is obvious to my students as well.

In schools where teachers can partner teach, it's terrific to identify the subject areas that are not your strengths to teach, your Kryptonite, and share those with teachers who DO find them interesting and appealing. In this situation, teachers benefit as well as the students. However, this type of arrangement is not readily available in all schools nor accepted by all school districts or administration.

If you're a teacher who is expected to teach all subjects, it's important to identify those subjects that are not as appealing to you as others. Once you know what they are, you can search for strategies and lessons that excite you about the subject and find ways to make those strategies part of your instruction.

Task 2: Plan, Plan, & Implement

Once you've identified your teaching weaknesses, your Kryptonite, areas where you might fall short on enthusiasm, it's time to plan. Whether I like it or not, I generally have to devote more time and effort in planning for my least favorite subjects. These subjects usually require me to do a little research on the computer in order to see what other educators might be doing to teach similar concepts. Sometimes, I might need a refresher for a math or science concept I haven't taught before. As I log onto my computer, I often wonder, what did teachers do without Google.

The following outlines the procedure that I undertake to get started planning a unit to ensure engagement and intensity of instruction:

- Determine the end goal of the lesson/unit;
- Identify the targeted learning outcomes;
- Look at the material from your perspective;
- Ask yourself, what might make this topic interesting to me and my students?

One tool that I find helpful is to keep a 3-ring binder of additional materials I create for each lesson as part of the unit. I have found, over the years, that struggling students in particular, need more practice and reinforcement to learn concepts and skills than is usually provided in textbooks or workbooks. By keeping the binder, I am able to refer to it, year after year, and add to it as new resources are developed. Another benefit of the binder is that it allowed me to start with a few units each year, gradually building up the number of interesting lessons available rather than trying to do it all in one year.

I've also developed a highlighting system so that I know which resources are used each year. For example, one year, I might highlight pages at the top with a yellow highlighter. The next year, I might use a light green highlighter, and keep track of the year corresponding with each color in the front of the book. The advantage to using a highlighter is that it does not appear when making copies of the original document, yet it lets me know which years I used each unit.

Many teachers I know scan their plans and keep their supplemental resources in folders on their computer. This works great, especially for the younger teachers who have used little but electronic files for much of their lives. Personally, I like to see the “hard copy” as I flip through the binder. It helps me to visually plan for each day’s lesson and the unit in general. Use the method that works best for you.

In addition to the 3-ring binder, I have each student complete an “interactive journal”. They have one for math as well as language arts that they keep adding to throughout the year. Generally, each student ends up with two of each for each subject, as I like to start a new journal when we return from holiday break in January. (Note: Students create interactive journal by writing, drawing, or gluing pages into inexpensive spiral notebooks.)

As I teach each lesson, the students’ completed small group work and homework is glued into the interactive journal with the strategy/concept and date at the top. I am able to make notes on the individual pages of each student’s journal, as needed. For example, I might highlight one area where a student struggled specifically and write in the margin, “Completed with assistance, difficulty with regrouping”. I might place a star next to a section where a student was exceptionally strong, and make a note, such as “Deshawn was a whiz on this” or “Manipulatives were a great help for this section.”

Because I tell students that they don't have to remember the answer to everything, I often let the students utilize their interactive journals to complete assignments and when appropriate as a resource for taking assessments. I think it's important for students to realize that they can use what they have already learned to help them learn new ideas and concepts.

I particularly like the interactive journals because they become a source of pride for the students. They love flipping back through the pages, often marveling, "Wow! Look at all the work I've done!" The journals also provide great documentation and data when I am reflecting upon grades and preparing for conferences. When I'm having difficulty thinking of comments for the report card, I can flip through the journal and write a note about a particular strong point, or an area the student persevered. Parents also enjoy looking through them during parent/teacher conferences. They provide great evidence for parents to see where their child excels and where their child needs additional practice. It also helps to keep the conference on track and helps me remember exactly what we covered this year. If you are new teacher, this probably isn't an issue, but the more years you teach the more likely that 'senior moment' will happen at the worst possible time. This is especially embarrassing if you are nowhere near old enough to have senior moments.

As far as social studies, science, and writing, I like to have the students keep "mini interactive journals" which include vocabulary and individual topics, such as voting for social studies, the scientific method for science, and using transition words for writing as examples.

The students can't wait to take them home at the end of the school year. I try to keep one copy of each year's interactive journal, as there are usually one or two students who don't share the same enthusiasm as the rest to save them. If someone doesn't offer up their journal, it's easy enough to make a quick copy for future reference. Another idea is to have students stack their throw-aways on a table rather than throwing in the trash. Then you can easily take one that is being discarded without dumpster diving. These old interactive journals are also great as an example to show students the next year.

The final task that should not be overlooked when planning for curriculum is to reflect upon lessons as they are delivered. Make notations somewhere. (I usually add my notes to my 3-ring binder.) If a lesson was very engaging, make a note of it. If a lesson needed manipulatives that were not originally included in the lesson plan, but should be next time, make a note of it. If a lesson falls flat on its face, I make a note of it, but I also try to create a resource or plan an activity that might be more successful next time while the unit is still fresh in my mind.

When I am really stumped about how best to develop a lesson or unit, I talk with other teachers who teach the same unit and see what ideas they have. This is especially easy to do if you are part of a school district that has begun to implement Professional Learning Communities (PLCs).

Regardless of how you choose to record your reflections, just remember, reflecting on lessons, engagement, and academic achievement are just as important as the lesson planning.

THEIR STORIES

"Research on brain functioning informs us that there is great diversity, and unpredictability in how individuals take in stimulation from the environment, process it, and complete the function we call learning. Add a challenging student or two to the class roster, whose brain functions are obscured by layers of mistrust, disability, boredom, and neglect, and the success of our lesson plans is anything but a safe bet. …certain students need to see the meaning and purpose of each task in order to learn"

Benson, page 141

Identify the Dilemma

GUSTAVO: There's No Crying in Fractions

One year, I had a student in my 5th grade, special education class named Gustavo. He was a great kid with a wonderful family and a very supportive Mom. She attended conferences, returned phone calls, and sent notes when he didn't understand his homework.

Gustavo was also a good student. He worked hard and wanted to do his best at all tasks. Reading was difficult for him, and with English being his second language, he struggled with phonetics and decoding. However, he tried his best no matter what was asked.

I soon discovered he had a school-day kryptonite, too! It was starting a new math unit. Every time, since the beginning of the school year, whenever we were going to begin a new math unit, Gustavo would be sitting at his desk, pencil in hand, with wet, quiet tears streaming down his face.

When this first occurred, I thought something was tragically wrong. Considering math was in the afternoon after lunch and gym, I thought it was likely that some incident had occurred of which I was unaware. To my question of "What's wrong?" he just shook his head from side to side and continued to cry. I asked if he wanted to step out in the hall for a moment, and again, all I received was the side to side shake of his head. Not knowing what else to do, I began the lesson.

When the crying started up at the beginning of the third unit of math, I put two and two together and came up with five (no, just kidding). I realized that Gustavo was becoming upset about starting a new unit in math. Yes, it was true that he struggled with math; however, given a week or so into the lesson, he worked his way through the difficulty and was able to complete the math work with support.

When this realization came to me, I realized there had to be something I could do to ease his feelings of uneasiness about starting a new unit. I couldn't, with good conscience, allow him to feel so stressed for the entire school year.

I devised the following plan for Gustavo: About a week prior to the start of a new unit, I spoke with him privately about the unit we would be beginning the next week. I also called his Mom and let her know that we would be beginning a new unit; so she could remind him of it a few times before the actual day. On the first day of the unit, I provided Gustavo an outline of the math unit with a breakdown of what key concepts we would be covering each day. It also showed which homework pages/problems would be assigned (if any).

Lastly, each day after the whole group lesson, as students got started on independent work or small group work, I met with Gustavo for a few moments and discussed any questions he might have about what I had taught that day. I also took a couple of minutes to make a note or two on his note pages to show how he could use what he already knew from a previous unit to complete the work assigned for that day.

This approach seemed to help ease Gustavo's anxiety over starting a new unit and gave him confidence to ask questions and seek assistance when needed.

Most importantly, I reminded Gustavo of his past successes in understanding the math concepts he thought would be too difficult. I kept reminding him of those successes every time he felt anxious about new concepts until the end of the year. Gustavo's interactive journal was a great way to show him all the things he had already learned.

Years later, I ran into Gustavo at the local Wal-Mart with his mother. I saw her first in the produce section and waved hello. She waved back and motioned across the produce section for someone. A tall young man, who towered over me by a good six inches, walked over and smiled down at me. It was Gustavo! He stood there shyly smiling as his mother recited his many accomplishments since his 5th grade year, including making the high school honors band and being the first in his family to take advanced math classes in high school. I asked if he was too old to give his "old teacher" a hug, he smiled saying, "Thanks Ms. Adams, you made math bearable." Good for him!

Because I was able to identify the Gustavo's dilemma and provide support, he was able to experience success in spite of his fear of starting something new. Many students are struggling with issues at school. By identifying these difficulties and providing support, the students can get down to the business of learning!

Take Your Audience into Consideration

ENRIQUE: What's in a Name?

I've taught classes over the years whose differing ability levels would be enough to make a novice teacher cry. In the same 5th grade classroom, I've had students who had memorized all of their math facts (usually addition and subtraction up to 20 not to sound too advanced), and at the same time, students who were still utilizing their fingers to add and subtract one-digit numbers up to 10. To say that planning for instruction that was engaging and meaningful to all of the students was a challenge is an obvious understatement.

To make the process more difficult, even students who are struggling with math concepts, such as addition and subtraction of one-digit numbers, do not and will not complete worksheets or engage in manipulative instruction if it looks like it is for a "baby". So I could forget the notion of using photocopied practice pages from the 1st or 2nd grade math book for these 5th grade students because their street-smart radar can sniff those pages out and they'll land on the floor, if you're lucky.

One year, I was attempting to teach basic angles to a group of lower students. Without realizing the impact of what I was about to do, I held up my right arm and bent it at the elbow, pointing to the outline of the shape and saying, "This is a right angle." "Oh," they responded, as they looked around the small group table as their peers did the same. I thought to myself, this is great! They're engaged in angles. I continued with the instruction using whiteboards and straws, and whatever other interactive objects I could dream up; I felt confident that most of these students understood basic angle concepts, like the difference between a 90-degree right angle and a 45-degree acute angle.

A few days later, as we were reviewing at the small group table for an angle assessment the next day, the principal came by to drop off some papers. One of the students sitting at the table with me proudly announced that we had been studying angles and he had them all figured out. The principal raised an eyebrow and smiled saying, "Really? Tell me what you've learned." Enrique proudly stood up by the table, held up his right arm, bent it at the elbow and said, "This is a right angle." I smiled and feeling encouraged, without hesitation, he held up his left arm and bent it at the elbow proclaiming proudly, “And this is a left angle!” “Really? That’s … um … great!” the principal managed to say as she left the room covering her smiling mouth with her hand.

Later, when she saw me in the hallway, she couldn’t help but tease me, remarking that I was really making progress with Enrique. I smiled back and said, “Live and learn, right? Guess I’ll keep my body simulations to myself. Knowing that everyone knew math was not my best subject, I added, I swear I know there is no ‘left angle’!” I walked away to sound of her chuckling.

Continuing with the same class, a month or so later, we were working on identifying geometric shapes, such as octagon, hexagon, etc. The lower group was enjoying using a set of stencils I’d found to complete the class assignment learning to identify and label the shapes. The following week, after successfully reviewing the concepts, the students took the geometric shape assessment.

That evening, as I was grading the assessments, I was feeling hopeful as I made my way through the papers, finding that the majority of the students were doing very well. Imagine my surprise when I came to Enrique’s assessment and discovered he'd done very well on the front page using the stencil to draw the geometric shape above the word. Turning to page two I became confused as I read Enrique's answers, and then, highly entertained as I realized what he had done. On the page, I had drawn geometric shapes with a blank line beneath each shape and the word "name" below the lines. On this line, I expected the students to use the word bank above to write the name of the geometric shape on the line. Instead, on each line below the geometric shapes, Enrique had written very clearly "ENRIQUE". Yes, that is correct, he had written his own name under each shape!

Again, lesson learned the hard way. When working with students of lower cognitive abilities (and even some who are not lower in level), it's best to make your instruction very clear, leaving little room for misunderstanding, aka, my right-angle demonstration. Further, where I had expected the name of the shape, perhaps I should have expanded on my expectation and written "name the geometric shape" instead. Another alternative for Enrique's group might have been to put the words along one side of the paper and the geometric shape along the other and ask the students to match the shape with the word. This would have eliminated the mistake Enrique made, and also would have made it easier for those students who have difficulty copying from the word bank onto the line without becoming confused with where they are on the page.

The idea of this story is that math isn't just confusing to me. It can be confusing to students as well! It's important to take into mind the audience you are instructing, and utilize clear methods when delivering instruction, and even clearer guidelines when requesting answers on assessments.

Lastly, it's important as I've learned over the years, to maintain a sense of humor about teaching. Short of teaching students it's okay to walk in the middle of the highway, most teaching faux-pas can be remediated through re-teaching and re-assessing.

I've found it's helpful when I remember to take my audience into consideration. This is especially true when teaching at-risk students. Ask yourself: How explicit do my directions need to be? What is the vocabulary level of my students? What background knowledge does this group of students possess? These things will help determine how much support to plan for when teaching grade-level concepts.

Connect All Curriculum Areas to a Fun Topic

MIKEY: School's Boring

Mikey was a 7-year-old boy in my 2nd grade class, one year. Each day, he undertook a mission to make himself unlikeable. I've got to give him a great deal of credit because he took on this mission like a dedicated soldier headed into battle!

Regardless of the activity I had planned, he announced, "this is boring" and for every assignment we did, he exclaimed "this is stupid". Notwithstanding his average intelligence, he struggled to learn grade level concepts and lagged behind his peers in academic growth; I think this was mostly due to his unwillingness to engage in any lesson.

Most days, he'd arrive and have a meltdown before ever making it to his seat. Usually, it was the mere reminder that today was music or asking if he happened to bring his gym shoes for gym. To these innocuous declarations or questions, he'd throw himself on the floor, laying there motionless, and moaning that "school is stupid." Getting him to start the day without some amount of drama was a struggle … and that struggle continued all day, with all of his teachers, regardless of the subject being studied.

At times, it was excruciating for me to restrain myself from being drawn into a debate by his constant complaining and whining. Even with my best efforts, I'd often find myself exasperated and send him crying to his desk because all my labors at engaging him were not working. He insisted he was right about everything, even when the answer to the math equation was incorrect. He'd refuse assistance and push my hand away as I attempted to show him the mistake in his calculations.

One night at home, I determined that even though I couldn't discern the reasons why Mikey was the way he was, I couldn't allow him to continue to feel dejected and disengaged at school. I decided that, despite his efforts to make himself unlikeable, I was going to "kill him with kindness".

The next morning, I greeted him at the bus as always, but this day, I laid on a gigantic smile, informing him emphatically that today was going to be a great day! Not having had too many great days recently, he looked at me very doubtfully, but surprising to say, took my hand as we headed off to the classroom.

When he'd removed his coat, I refused to be drawn into his complaints about why kids couldn't bring toys to school and why school couldn't be fun like video games. Instead, I announced that I needed a "big squeeze" to get the day started as I was feeling a little bit tired. Much to my surprise, he ran over to give me a "big squeeze" around the waist and headed off to his seat.

That day, I made a point to offer him many compliments when I found him doing something he was supposed to be doing. I attempted to field his questions that were unrelated to the topic at hand and managed to get him interested in an activity during reading. He was always talking about storms, so I'd planned a VENN diagram activity for him to compare a "Hurricane" to a "Tornado". So far, so good, right?

At the end of the day, I was exhausted! It wasn't easy smiling at a kid who was insisting that his way of doing math (however wrong) was really the right way to do it! It wasn't easy smiling when he stated for the hundredth time that "school was boring" and "why couldn't I be as fun as a video game". But smile I did, and Mikey made it through the day with the fewest meltdowns since returning from Christmas break.

As I put him on the bus for the weekend, I reminded him that Monday was going to be another great day, and I was rewarded with one of his rare smiles and two thumbs up.

That afternoon, as I drove home from school, I reminded myself that I'd better rest up because my campaign to love Mikey, even while he battled, required a great deal of energy, patience, and resistance to his efforts to suck me into his negative perception of school.

Learning that Mikey wanted to be a storm chaser when he grew up was a wonderful discovery! Little did I know until he drew a picture of tornado and said, "This is an F-5 tornado! And that's you spinning around in the middle of it." It wasn't just a statement of how much he was detesting me at the moment, it was also a declaration of a subject he was willing to invest some time into!

Laugh as you most likely did at the mental image of his drawing, but let me tell you, educating Mikey was hard work. He hated anything he perceived as being too difficult for him (even if it wasn't) and he wanted nothing to do with anything that required cooperative learning (as no one knew as much as he did about everything – according to Mikey). So, keeping him moving forward in instruction often became a battle; some of which I wouldn't be too proud to describe. Not sure who won those battles most of the time, except I can be certain that Mikey wasn't enjoying learning and a great deal of the time, I wasn't enjoying instructing him. That's why Storm Chasing became a subject I could throw out and immediately he would raise his head up from his desk with his ears to attention. Remembering to drop in a little video or story about the weather helped get Mikey and me through most of his second grade year.

Late spring, as it was getting close to the end of the school year, Mikey was having a particularly difficult day. It seems he "ran out of time" during art and didn't get to finish his painting and his shoelace was broken and he wished he had his tablet … and let the complaints continue. Needless to say, I was attempting to talk with him and calm him down in hopes of enticing him to do a few math problems before he went home. At the time, he was laying in the doorway whining.

Attempting to distract him, I said excitedly, "Guess what? We're going to be Storm Chasers during summer school! Won't that be great? We can measure rain and study storms! What do you think?" With that question, he threw back his head and began crying, "I can't do that! My Mom won't let me get that close to tornadoes!" he wailed. I stated calmly, "No, we're not actually going to be Storm Chasers. We're going to pretend and study those things from the classroom. That'll be okay with your Mom, don't you think?" He wailed even louder, "Well, what's the fun in that?!" At that juncture, I gave up for the day, getting the feeling that there was nothing I would say today that would get Mikey to the math table.

Fast-forward to summer school, I planned an entire month-long unit on Storm Chasers which included reading, writing, math, science, and social studies. And even art! I would not be completely truthful if I said that Mikey was engaged and excited every day of summer school; however, I will tell you he was engaged more days than he was not!

It was fun for me, as the teacher, to plan for a unit that I knew would be interesting to my students, and to know that I would be covering all curriculum requirements with that one unit. No additional planning for math. And being able to incorporate art into the unit was a special treat for the students because the regular school year instructional pace allows for little time to do art in the classroom.

As for Mikey, he ran to the bus the last day of summer school smiling widely as he climbed on the bus with his hand-made rain gauge and painting of his F-5 tornado!

This month-long unit can be found in additional resources at the end of this book.

(A portion of this story appeared in Don't Look Too Closely, published by Belinda Adams, 2018.)

By connecting all curriculum areas to a fun topic you know your students find interesting could be the answer for the student you have been struggling with all year. During the regular school year, it might be easier to do a 2 or 3-week unit. Getting to know your students and their interests may offer great ideas for engaging lessons.

Incorporate Emotion to Increase Engagement in Social Studies Lesson

Jake: Mrs. Adams, You've Been Discriminated!

It's Black History Month. Happens the same time every year, and somehow, we're supposed to make it meaningful, new, and interesting. I placed words on the board: Discrimination, Racism, Segregation.

Having looped with this group of students, I knew I couldn't open up the Black History Month folder on my laptop. I needed a new bag of tricks. I had to find a way to drive home a point that few of them recognized; how many thanks we needed to give to those timeless heroes like Ruby Bridges, Harriet Tubman, Rosa Parks, and Martin Luther King Jr.

My instruction began like the year before: "Let's talk about what we remember about these influential African Americans." Yada. Yada. Yada. I could see the looks on their faces. They weren't being disrespectful, but they certainly weren't engaged in the discussion, they thought they had heard it all before. That is, until I told them I had been discriminated against, too! "What?!" They shouted as they sat up in their seats, indignant that someone could discriminate against their teacher of two years and astounded that something like that could happen to this blonde-haired, blue-eyed teacher.

"Yep," I said, "that's right, I've been judged by others who didn't know me, based on how I look." I went on to explain how I felt those first few weeks teaching in the low-income, mostly Hispanic elementary school where I'd started teaching. I described the first morning I went out to greet my first-graders with a wide smile, only to be greeted by their parents with looks of doubt and pessimism. I told my students how I felt like I could read their minds as they looked me up and down from toe to head, wondering how in the world this white woman was going to teach their children anything! I continued to explain how the same experience had been repeated when I moved to another school to teach 5th grade. I described how it felt to be scrutinized by the African American parents who'd come to drop off their child's school supplies. Again, I felt their skepticism that I'd be able to do their child one bit of good.

My students just sat there, their mouths agape and I could see them playing the scene out in their head. We talked about what choices I had during those moments. I said, "Well, I could have sought revenge and made little Jose's life miserable because his mom doubted my capability." Still, no words came out of their mouths as they waited to see if indeed I had made little Jose's life miserable. I finished, "Instead, I chose to prove them wrong! I decided I'd do my best to teach their child, regardless of their disbelief in my ability. And I'd like to think that they changed how they felt about me as they watched their child learn to read."

"Way to go, Mrs. Adams," they shouted while pumping their fists up in the air. I smiled right back. I asked if they had the same options as I had when faced with situations where they felt judged? "Sure, why not?" they answered, filled with optimism that it really was as easy as I'd made it sound.

Without passing out a single paper, or assigning one homework assignment, I had taught my students a lesson about discrimination that I felt they would hold onto for longer than any story from a book. In addition, they were ready to tackle other stories of discrimination, such as Ruby Bridges, Rosa Parks, Martin Luther King Jr., and Jackie Robinson.

I know it's not as easy as I'd made it sound that day to enlighten students to a concept as complex as discrimination, but I'd like to think that day I may have planted a seed that caused a few of them to reconsider their notions of discrimination, and to reconstruct some options of how they respond to these situations. I wanted them to believe they are really the ones in charge of how situations like those conclude. Students also learned that those same topics which appear again and again throughout school curriculum, such as discrimination, don't have to be mere repeats of the year before. They can use what they've learned to discover more and become active participants in what they choose to discover and how deep they are willing to take the material. When students are emotionally engaged they are more in tune with the lesson and put more effort into learning activities.

(A portion of this story appeared in Can You See Me? published by Belinda Adams, 2018.)

Incorporating emotions to increase engagement is easiest when you rally students to a cause. Try a personal story or a current event in the news to engage emotions.

Nourish the Spirit, As Well as the Mind

Emily: Those Other Teachers

One year, my partner teacher and I had a student named Emily. She was a 5th grade, special education student who had great potential. She'd lost a great deal of academic time due to a degenerative health condition in her legs that had resulted in several surgeries and for which she continued to experience a great deal of discomfort.

Not that you'd ever hear much complaining out of her. However, if one watched her closely, you might see her wince when she got up and down from her chair. Those days, my partner teacher and I would quietly ask, "Do you need to go to the nurse for one of your pain pills?" She'd usually nod affirmatively and go down the hallway with her head held high.

Due to her time out of the classroom, Emily struggled a great deal at the onset of school with reading and math. However, with some individualized support during class and her lunchtime, Emily was making great strides in both subjects because she was a hard worker and wanted to 'get smarter' in her words!

We worked out a plan where Emily would try everything first. If she didn't understand, she would come see me to get some extra help. It was fortunate for her that Emily attended the Boys and Girls Club after school. This made it convenient for her to stop by after school if she needed more help than I or my partner teacher could provide in a few minutes available during class. With this one-on-one help and Emily's determination to be successful, she made tremendous progress. I'd like to say it was because of the wonderful strategies we implemented that were just not available in the general education classroom. However, that was not the case. Emily was a student who just needed a little extra support to be successful in the general education classroom. Knowing that, for her 6th grade year, we placed her in the general education classroom with a teacher's aide for extra support.

About a month into the school year, I saw Emily in the hall with her bright smile. She said "hi" and asked if I had a few minutes to talk. It was during my planning period, so I said I would love to talk. She proceeded to tell me how she loved the big class, especially her literature class. Then I noticed her smile had disappeared and she said to me, "I'm not sure I'm smart enough to be in the big math class."

When I asked her why, she said she needed help and asked if I would tutor her after school, saying she just couldn't understand how to do the 6th grade math. We agreed she would come by after school on Tuesdays. Before the 1st Tuesday arrived, I stopped by to see Emily's math teacher to ask what they were working on, letting her know that I would be tutoring Emily. I was shocked when the teacher refused to share the information saying, it wouldn't be fair if Emily got extra help. She continued to tell me that they were using a problem-based discovery method. They did not explain to the students how to do the problems, rather gave them problems and the students figured it out in their groups. Evidently, Emily's group was doing well, so she said Emily did not need any more help, that the group would help her.

I was floored! I didn't know what to say. My philosophy had always been to provide a student with strategies to help them be successful whenever I could. If one strategy or teaching method was not successful for a particular student, I had no problem looking for a method that was helpful. Not only that, in all my years of teaching, I had never had a student ask me for tutoring. Sure, I have had some parents ask, but never a student. This was definitely a behavior I wanted to encourage.

So, I had Emily bring her journal (they were not allowed to take math books out of the room). You probably won't be surprised, but I didn't understand *that* either. I took her to the copy room and copied the pages and told her to come back next week. After she left, I went to a friend who loves teaching math and asked for help. Turns out, it wasn't something I didn't know, I just didn't recognize it in the discovery format.

When Emily came back on Tuesday, I proceeded to show her how I had learned it in school. She smiled and said "thanks," it makes so much more sense now. We agreed that whenever a new topic was introduced, if she didn't understand she would stop by my room, and either my partner teacher or I would figure it out with her.

The bottom line, sometimes when teaching math and one or two kids just don't get it, you need to teach them a different way, not only the way it is in the teacher's manual. Usually, if you watch a child work through a problem, you can see where the student is making the error. Sometimes, it is just prerequisite knowledge the student is missing; other times the student needs different strategies. One size never fits all. Using different strategies is important for the child's spirit. Failure is taken personally by children, crushing their spirit and reducing the chance of future success as students internalize the message that they are a failure. Don't let this happen without trying other strategies.

Sadly, this form of "discovery" instruction and independent work generally isn't successful for special education students, especially those recently joining an inclusion classroom of general education students. Further, we know many general education students for whom this type of instruction does not work either, especially in a building which consists primarily of low-income students who are struggling with grade level curriculum.

Nourishing the spirit as well as the mind is often an essential strategy for at-risk students. It's important for us, as educators, to remember that some students don't learn "naturally" without any instruction! They need to be educated explicitly, re-taught, shown how-to through modeling, and given an opportunity to ask questions and receive immediate feedback. Helping a child experience success through the use of multiple strategies is the best way to heal a spirit broken by failure.

"What is important is that my students think that their beautiful minds are stimulated into activity, that they feel the pleasure of engagement, and it's worth it, and it encourages them to live. To this end, the goal of every class is to inspire lifelong learning."

Benson, page 143

Imagining Future Success Fuels Focus

Demarco: Let's Get Rich Together

After about my first or second year of teaching, it became apparent to me that the last month of school can be extremely taxing, both for teachers and students. Worn out and tired of one another, makes not seeing any future school holidays on the calendar after Good Friday very discouraging. This causes moodiness and a reluctance to get out supplies. Yes, I'm talking about the teacher, but the students are generally feeling that way, too.

Keeping this knowledge in mind, a few years ago I created an end of the year unit that "I'd saved especially for this class because I know you'll take it and run with it." I call it "The Entrepreneur in You". We spend about a week researching, writing, and discussing 10 young entrepreneurs who have made hundreds of thousands of dollars or more, creating everything from sock puppets to greeting cards to homemade soap.

The students got really excited reading about these young entrepreneurs! I think their enthusiasm has a great deal to do with first, discovering that kids can be rich, and second, showing that there may be a future for them outside of the one they may be envisioning now.

During the second week, the students decide on a business they would like to start. Students research other businesses like the one they want to start, name it, create a “mini business plan”, and create an advertisement to tell others about their new business. You can almost feel the excitement as the students get more and more involved in their businesses, discussing it with one another and making plans for the future.

The third and last week of the unit, the students design and present their advertisements to the class. Over the years, it’s become much more fun, as technology opens up many avenues for designing their advertisements, including Power Points, videos, songs, and more. After each student presents his or her adverting plan, the class provides positive feedback and suggests improvements to each student’s business and advertising plans as long as the comments are of a productive nature.

When I have higher students, I can take this unit even further. We can utilize math concepts to determine possible supply needs and budgets to get started. Students can have the opportunity to price their products or services, and even go as far as calculating how much money they might make. Anytime, we can link math to the unit, but determining how a certain math skill we are learning would be needed by the kids in their businesses. Adding science to the mix, we can determine what environmental impact their business might have and what can we do to make things better.

I love this unit. I look forward to it each year. I'm often surprised by the creative ideas the students come up with, and the enthusiasm with which they tackle the project. What makes this unit very appealing to me, and should be to other educators as well, is the ability to add to and minimize the tasks to make the unit expectations appropriate to the skill levels of the students and the requirements of the curriculum.

You might ask yourself that if this specific unit doesn't fit with your curriculum, is there another topic you can introduce that will generate this enthusiasm? Although, I have to stress: what kid doesn't dream of being rich? Whether it's as a professional athlete or rapper or actor, most dream of having lots of money someday. The idea is to teach while kids are working hard because they like what they are doing. One thing I have learned over the years, no matter how important the skill or topic I am teaching, if students do not engage in the material, any learning that takes place is minimal.

Jensen (2013) talks about developing engaging instruction and defines that as any instructional plan that gets students to participate emotionally, cognitively and behaviorally. Further, he says that engagement happens when teachers use methods for instruction that stimulate, motivate, and activate their students. He states, "When the class's physical and emotional energy is high and flush with hope and optimism, students work harder and enjoy the learning more" (p. 36).

(A portion of this story appeared in Can You See Me, published by Belinda Adams, 2018.)

Imagining future success to find focus can be done through any unit that looks at future successes. It can be something as simple as discovering what students want to be when they grow up. That can be a great starting point for planning a unit.

Sometimes You Just Have to Add the Extra

Daniel: Math Rocks

Daniel was one of those students who just grows on the teacher over time. Because "love" wasn't the first emotion I experienced while having interactions with Daniel at the beginning of the school year.

On the first day of school, he was compliant and willing. However, two short weeks into the school year, he turned sullen and resentful. And it didn't take me long to figure out why. School was difficult for Danny. Every subject. Every day.

He seldom showed a smile, unless it was time for lunch or recess. He never turned in homework and rarely completed his classwork. In the beginning he always had an excuse for why his homework wasn't completed, but before long, he would just say with little emotion, 'I don't get it."

I recall that was the year the district started their monthly timed tests for math facts. They had determined that this form of assessment was to be the measurement we were using to measure student growth in math.

If you know me, I'm sure you can imagine that I hated those timed tests, almost as much as the kids hated them. True, I had a couple students who were motivated by the tests; however, only one of the students made progress with this motivation. These students were up against many odds in their day-to-day life, with poverty being at the top of the list, and learning their math facts was certainly not a priority for any of them. Generally speaking, the timed math tests were looked upon by me as an inconvenience and to them, just another opportunity to show the world they were a failure in math.

I checked around with several other teachers who also didn't seem to be seeing academic growth with the timed tests till I discovered a program called Rocket Math. We started doing it 3 days a week. It is a supplemental program that requires less than 5 minutes of instructional time per day. This approach solved the problem of students not knowing how to learn their math facts. In addition, students benefitted because Rocket Math only adds 3 or 4 new facts for each level, making it do-able, even for those who struggle with rote memory. As we used this program, I had a few more students who were beginning to show minimal improvement. But motivation was still lacking for most of the class.

After brainstorming with a couple other teachers in the building, a plan came to fruition. The following Monday, when it was time for Rocket Math, I told the kids I wanted to discuss something with them. I asked, "Do you know why I am a teacher?" My seemingly innocuous question was met with silence, and a few glances thrown amongst the students. I answered my own question by saying, "I'm a teacher because I love school." More silence. For most of them, school had been a place for failure and punishments. I continued that, "I love learning new stuff, and I learn something new all the time. And even though some of you might doubt it, I love working with kids." More silence, but now I had their full attention. "But I also do this job because I get paid. Everyone needs money to live; and I am fortunate enough that I get to do a job that I love that also pays me money to do it!" One student yelled, "How much?" Another seconded, "Yeah, how much you get paid?" I said easily, "Enough that I can afford to buy the things that I want, like new shoes and going out to dinner when I want." Now, they were really interested.

I continued, "So, if I get paid, why shouldn't you get paid for your work? Starting today, you will get $2 in Bulldog Bucks for every level you pass in Rocket Math! And it gets better! You will still get $1 in Bulldog Bucks even if you don't pass the lesson. All you have to do is complete the practice sheet." "Sweet!" someone whispered, knowing full well the great candy and other goodies they could purchase with their Bulldog Bucks. I concluded by adding that our class would hold a monthly auction of "goodies" for items they could purchase with their Bulldog Bucks.

(A description of how to set up an auction without going bankrupt can be found in the resources section of this book.)

Now back to the story of Danny. As it turned out, memorization of rote facts was the one area of academics that Danny actually excelled in! As part of the program, I posted the students who were at the highest level. Not wanting to embarrass anyone, I didn't post all of the students, only the top 5 students. I encouraged the rest of the class telling them each one of them had the ability to make it to the top 5. (Really, my math isn't that bad. As students complete program, they are no longer counted.)

With this approach, Danny made his way to the top of the list week after week. As such, he found a new standing in class among his peers, experiencing success at school for probably the first time in his life. The students who initially thought Danny was “dumb” were soon calling him the “math whiz.” Honestly, this approach didn’t make Danny a math whiz as it’s hard to make up for years of academic inactivity. However, this strategy did give him the confidence to work diligently at his math facts, and also gave him some confidence in his ability to learn math concepts.

Most importantly, the success gave him back the smile we hadn’t seen since the first weeks of school. And even though I’ll admit that rote memorization is not my favorite approach to school, it made me realize that it could be a bright spot for a student! Even the kids who struggled could get rewarded! For Danny, being the champ of Rocket Math made all the difference!

Sometimes you have to add the extra to get kids excited. Whether it is an auction, a class party, or extra recess, make sure that all students have the ability to participate, using various goals for each student if necessary.

Even a smart kid "with school problems needs coordination of support, evaluation of strengths and weaknesses, anticipation of curriculum demands, and accommodation of learning style."

Administrators "need classroom teachers who accept atypical learning styles and are willing to use a variety of materials and teaching techniques."

Vail, pg. 16-17

Consider Preferential Practices & Eliminate Power Struggles

JOEY: Math is Scary!

Joey was a boy after my heart from the first day he crawled under his desk, and stated quite matter-of-factly, "Maff scares me." What I wanted to whisper was, "Maff scares me, too, kiddo! Let's get outta here!" However, as I'm sure we'd all agree. That would have been very inappropriate, especially for a first year teacher.

In addition, I was the leader of the class! I was supposed to set an example for the rest of the students to follow. What example would I be setting if I joined little Joey under his desk to await lunchtime?

Although Joey had a fond spot in my heart for our shared distrust of math, Joey brought some difficulty to my instruction as well. He was a student who, without really even trying, made my job of teaching him more difficult for a few reasons. He frequently missed school or was tardy at least 3 out of the 4 days he attended school, he didn't speak often, and rarely responded with more than a one-word answer, he showed little to no emotion on his face (unless it was recess), and he hid under his desk every day when it was time for math!"

If you're an educator, you know how hindering frequent absences are, especially during the primary years when skills build a foundation like building blocks, one upon the other with each passing day. And Joey, already behind in his 1st grade class due to missing more than half of his Kindergarten year, couldn't afford to miss school as often as he did.

Out of frustration for Joey's' situation, I went searching for information. Seeking out resources in the form of a few other teachers in the building, I was able to piece together some facts about my little absentee boy.

Joey lived with his mother who spoke only Spanish and his father who spoke English and a little Spanish and his older brother who spoke both Spanish and English. His mother worked the night shift and his father worked the day shift, and neither shift was conducive to getting Joey to school on time. It was the responsibility of his high school brother to get himself up and get Joey to school each day.

Not an easy task when the big brother's absenteeism was just as high as Joey's. A tenuous situation to be sure.

Working with Mom and Dad, the social worker and I came up with a possible solution to get Joey to school, even if he was late. His parents were stunned to hear that his first-grade teacher wasn't going to lecture about Joey being tardy for school. It seems they'd already heard enough lectures about tardiness during Joey's Kindergarten year. I answered, "Are you kidding? I want Joey at school each day, and I don't care if he doesn't get here until lunch; just get him here!"

With that understanding, Joey's attendance improved. Joey began to make some academic progress and I got excited when I started to see his potential showing. I did additional planning that I thought might engage him in our instruction and did everything short of standing on my head to solicit a smile. Until "maff" no longer scared him, I passed him a lapboard with his math paper and pencil and he did his math under his desk. Little by little, a relationship of trust was built.

Another preferential treatment that Joey received was that he often complained that the din of the other 30 students in the classroom made it difficult to concentrate. With that in mind, I talked with the media staff and was able to obtain a set of headphones used for computers. At first, when I asked Joey if he wanted to wear them to help him concentrate, he looked at me questionably as if he was sure I'd finally lost my mind completely. "Where am I gonna plug them in?" he muttered. I said, "Nowhere. Just try them. If it doesn't help with the noise in the classroom, you can give them back." Skeptically, he looked at me, but put them on anyway.

Joey never said that day if the headphones helped him concentrate or not. He just solemnly handed them back to me after math on his way to gym. I decided not to mention them again; however, the next day, when it was class work time for math, he approached and pointed to the headphones I had lying on my desk. Without a word, I handed them to him and he headed to his desk, put on the headphones, and went to work (under his desk, of course!).

I've kept track of Joey over the years. When I saw him in a 5th grade classroom, he still had a cautious smile but came to school every day. He had friends and had learned to excel at academics, even "maff".

Kauffman echoes the positive effects of the instructional measures I took in the classroom by saying, "Preferential interventions by the teacher sometimes solve problems and avoid referral to special education" (Kauffman, 2005, p. 26).

Did I think, at the time, that providing Joey with a lapboard with his math paper and pencil under his desk was going to help him get past his fear of "maff"? Absolutely not! At the time, I was looking for a way to get this kid to do his math, while at the same time, avoiding a power struggle that rarely results in success for the child or the teacher.

(A portion of this story appeared in Don't Look Too Closely, published by Belinda Adams, 2018.)

Preferential treatment can often avoid power struggles. When other students have questioned, "Hey, how come I don't get to wear headphones?, I've used a variety of answers from everything to "I can get you a pair if you really want to wear them" to "that's something Joey needs to do his work, but you do your work awesomely without them." Students are usually placated with very simple answers to questions like that.

"Sometimes, what we do intuitively, as educators, to facilitate the process of learning using the path of least resistance is the best approach. One might even wonder if it is fair to give a child preferential treatment. I always remind

myself, life is not a level playing field; certainly not for the Joeys of the world. Anytime it is within our power to do something for a child who is struggling, we should do it without hesitation" (Adams, 2018, pg. 52-53).

"Because failure is easy to see, unenlightened adults may ignore the student's talent,

fostering a negative self-concept by focusing on what he cannot do."

Vail, page 1

Sometimes You Have to Start from Scratch

ANTHONY: Is that 13 or 31?

When I met Anthony after holiday break when he transferred to my class, I was amazed to discover that he was a first-grade student who didn't recognize any of the letters in his name or a single number in isolation. Yes, he could very loudly recite his A, B, C's and "count" up to the number twenty (if you wanted to forget there was a number 13, that is!).

However, when shown the letter "N," he couldn't even hazard a guess, even if I told him, "the letter is in your name." And when shown the number 7, again, he had no idea. Nor could he actually count 7 objects.

Scary, right? How does a student reach the age of 7 without any age appropriate skills? Sadly, he'd moved from school to school due to his family situation. To make matters more complicated for him, he'd been diagnosed with "special needs," adding further reason for some to excuse the fact that he had not managed to learn a single skill since starting school in Kindergarten.

One strategy I engaged for Anthony was using clear numbers and a 100's chart (showing only up to 20) for counting with one-to-one correspondence and number recognition. Another was matching letters on a page with magnet letters. Both of these were done as a daily warm up activity before beginning any lesson. While Anthony could say the numbers to 20, he could not count 10 objects.

At the onset of our process for teaching Anthony, I'll be honest and say that his "warm-up" was his entire lesson. Why? He found many, many reasons not to do the work! He cried. He pouted. He asked to go to the bathroom. He lost his pencil. And once, he even wet his pants.

Sound extreme to avoid math? Not to Anthony. You see, he'd spent the last year and a half in school avoiding doing all work and learning very little in the process. When the task of learning to write your own name is cumbersome, it's amazing what some students are willing to do in order to avoid tasks they deem beyond difficult.

Sticking to the plan, my para and I were able to reduce the amount of time that Anthony spent each day complaining, and slowly but surely, increase the amount of time he spent *actually* doing math and learning the alphabet!

Fortunately for us, and Anthony too, as time passed learning his numbers and copying letters of the alphabet came a bit more easily to him. Consequently, he was able to focus more attention on those tasks with much less drama.

After a couple of months, we were able to introduce Anthony to actual class work, and less rote skills, and his interest level improved. Subsequently, so did his academic achievement.

Another exciting thing happened when we discovered Anthony's love for the i-Pad. You see, his family couldn't afford luxuries like an i-Pad, and he was infatuated with the device. It soon became clear that my para and I could use the i-Pad as an incentive to motivate Anthony to complete the tasks he found less appealing.

We'd find ourselves smiling behind our hands when Anthony would become so excited yelling, "Is it time for Cool Math Games? Is It? Is it?" You see, this was a website he was allowed to use on the i-Pad, and to Anthony, math was no longer a chore when he was able to play games on the i-Pad.

Soon, we were matching him minute for minute with work completed and time on the tablet. It turned out to be a great motivator for Anthony! And despite his best efforts otherwise, Anthony was learning for the first time in his 7 years of life!

Sometimes you just have to start at the beginning. It is not uncommon for at-risk students to be so far behind that they are unable to be successful with grade-level tasks and to shut-down, refusing to engage in learning activities. In these cases, rather than punish the student for behavior we must teach the skills and concepts that were not learned with the goal of closing the achievement gap.

Wrapping Up

At the end of the day, when all has been taught and erased from the board, there are a few concepts to bear in mind. Teachers are good at teaching … we're better at helping our students make exceptional academic growth when we are passionate and engaged with the material we are teaching. Not as easy implemented as written, I know.

Hopefully, with the application of the strategies in this book (and some I anticipate you've envisioned on your own after reading this book), you will identify methods of teaching all subject matter, even those subjects that don't make your creative juices flow. Because if you're like me, my creative juice will never "flow" for math … at best, I might be able to get it to "ooze" like sap from a maple tree!

However, with a little research, a great deal of planning, and even more time spent instructing and reflecting, we can engage our students in all subject areas. Find ways to make lessons interesting to you, and more than likely, if you enjoyed planning and teaching it, your students will enjoy learning it. (Recall, I said, more than likely.)

There have been plenty of times when my instructional "aha" moments have fallen flat on their faces in the classroom, and I've had to retreat home and plan again. Other times, I've asked my husband or son, "Hey, would you like doing this?" If they look at me skeptically or worse yet, say something like, "What's my other option, a root canal?" I know I've still got some creative planning to do!

In addition, don't forget to take into account that every class each year is different than the previous year. It's not enough to dust off that interactive journal you kept from last year or pull out your teaching folder for fractions and head to the copy machine. You've got to turn a keen eye to the students in your classroom this year and make adjustments accordingly. It is important to engage every student and one way is to look for strategies like the ones in the kids' stories to see what each student needs to be successful.

I LOVE the school years when I have a group of kids who seem to just be excited about everything! They love counting money … "Yay!" They think practicing elapsed time is fun … "Yay!" They make my job so much easier, and consequently, I also get the privilege of watching their academic growth explode.

However, there are still those years (or those few students) who can't seem to find the fun in anything you plan. Regardless of the fact we'd like to exile them to the corner of the classroom, we've got to plan for those classes and those students as well. We have to take into consideration their individual needs and interests if we expect to bring them along on the engagement path to higher level learning.

Hopefully, this book has shown you that doing just that isn't impossible … it just takes time and effort. And that time and effort spent up front planning for engaging all students (and yourself) in the educational process pays off in higher student engagement, greater academic gains, and yes, fewer behavior issues in the classroom. I'd like to tell you that my aversion to teaching math – my kryptonite – has decreased over the years. I'd be lying, and too many who might read this book would be happy to call me out on it! However, I will tell you, when I am prepared with a great lesson that I find engaging, regardless of the subject matter I am teaching, it's easier to instruct, field questions, and muster excitement for the subject material.

Now when my student says excitedly with a smile, "Mrs. Adams, I didn't know you loved tornadoes, too!" I can honestly say, "You know what? They're growing on me!"

Resources

The Storm Chaser Unit

The purpose of the Storm Chaser Unit was to plan for an entire month of summer school with one single topic and include all curriculum areas. As you read previously with Mikey, he wants to be a Storm Chaser when he grows up. Therefore, I was thinking it would be an opportunity to engage a student who struggled daily with engagement and focus. I knew the other students would enjoy the unit because who doesn't want to learn about tornados and hurricanes and lightning?

For the unit, I created a small binder and separated the individual curriculum subjects with a colored piece of paper. That way, we could easily move from the "reading" section to the "math" section just by looking for the colored sheets of paper.

The Unit:

1) READING: Learning about kinds of weather: Each day, the students started with a different one-page, quick read about the type of weather we would be studying that day (lightning, thunder, rain, tsunamis, etc.).
   Students also read quick reads I re-typed about different kinds of Storm Chasers, the education or knowledge needed to be a storm chaser, and the "gear" required to be a storm chaser. (I re-typed the passages because I was unable to find a short passage at the primary reading level.)

2) MATH: Math problems were made "weather friendly" just by placing them into clouds and sun. In addition, students worked on concepts of measuring using real and paper rain gauges. Students also worked on reading larger numbers by reading real and paper thermometers. (These were 1st and 2nd grade students.) Students incorporated higher-level math by determining the difference between last Monday's temperature versus the current Monday's temperature, for example. Mikey was so excited when he was able to determine the wind difference between and F4 and an F5 tornado! (I won't tell you the answer; you'll have to look it up!)
3) SCIENCE: Students studied various landforms that are caused by weather. For example, did you know that an earthquake can cause a volcano to erupt? Students studied various kinds of rock, such as the different forms of rock that are part of a volcano. This also led to an investigation of different kinds of rock formations and layers of rock, etc.
4) SOCIAL STUDIES: Students studied various places around the globe that had experienced different types of weather and why.
5) ART: Student projects included: painting a picture of their favorite type of clouds with a short explanation; creating a volcano out of painted coffee filters and tissue paper; painting their favorite kind of weather; and cutting out and coloring their own personal storm chaser with required gear, such as a flashlight, computer, extra batteries, etc.

6) WRITING: Students participated in the entire writing process, using VENN diagrams to compare types of weather, preparing concept maps to write a basic paragraph about a certain type of weather, and finally, the culmination of all of their writing efforts was putting together a paper about what they would like best about being storm chaser and the specific types of weather they would like to "chase" as well as a list of the gear they would need to be successful at their job. The students shared their papers aloud in a "presentation" format and even answered questions from fellow students about what they had written. (For older students, consider having them create a book using the various writing assignments.)

The presentation was a great ending point for the unit as students took great pride in sharing what they had learned throughout the Storm Chaser Unit. They were excited to share their "favorite" weather. In addition, it gave all students an opportunity to speak in front of their peers and "be an expert" on the subject.

How to Set Up a Classroom Auction

The idea of a classroom auction is to provide external motivation for students who are having difficulty buying into the idea that education is valuable. In the process, it would be quite easy to spend yourself into the poor house. So, if you are like most of us and money is not in unlimited supply at your house, here are a few ideas.

First of all, candy is great. Because you don't want to give your students unlimited access to something with no health benefits, the idea is to stay small. Buy the 6-packs of snack size of Twix, M&Ms, etc. In the beginning, you can start with just 2 kinds of candy. Also, do not auction 6 or 12 little candy bars. Only put out a few, this drives the price up. I have had students pay $25 in Bulldog Bucks for a snack size candy bar. Another idea along this line is little juice boxes. The Dollar Store will have packages for $1. Again, only have a couple so the price is forced up. The little bags of chips or crackers are another inexpensive item. The Dollar Store has packs of 3 or 4 for one dollar. *Be sure that you are aware of allergies whenever you are including food product.*

The next idea is school supplies. During the summer when things are inexpensive, stock up on them. Students love things like novelty erasers or pens. Little notebooks are also popular. You can get 3 or 4 in a pack at the Dollar Store. Pencil sharpeners with covers, crayons, markers, and cute pencils are all items that students like.

Toys are near the top of the list in popularity. Clearance sales can have a lot of bargains. Look for party favors in packs that can be opened and auctioned off individually. After holiday sales are the perfect time to stock up. Often, holiday merchandise isn't specific to the holiday.

Finally, don't forget coupons, which do not cost you anything. Everyone loves the "homework coupon" (one assignment that doesn't need to be completed). Just make sure students know that it is not valid on multiple assignments; just one. Our class always had a "PAT coupon" (preferred activity time) that could be used for extra computer time or time to draw. The all-time favorite was a coupon to use the "teacher's chair" for a period.

The idea is to keep it simple and inexpensive. Don't feel you have to have a huge store of items to start with. Start small and build your collection. If you go places where they are giving out things with advertisements, like balls or other toys be sure to put them in your auction box. If your kids get toys in kids' meals that they don't want, in the auction box they go. You don't have kids? get them from nieces, nephews, and friends' kids. Let your friends know. Some people just love to find bargains and will be happy to have someone to give their 'finds' to.

One rule I always had was students could not buy items for a set price. The auction itself it part of the novelty. School stores are old. Plus, the auction causes the prices to be driven up and you can use that to show cause and effect, inflation, and other real-life concepts.

We usually had an auction every 4 to 6 weeks, but you can do what works for your schedule.

Bibliography

Adams, Belinda. Don't Look Too Closely: What Children of Trauma are Hiding & How You Can be the Difference for Them. 2018. Anchor Book Press, Palatine, IL.

Benson, Jeffrey. Hanging In: Strategies for Teaching the Students Who Challenge Us Most. 2014 ASCD, Alexandria, VA 22311.

Jensen, Eric. Engaging Students with Poverty in Mind. 2013. ACSD

Kauffman, James M. *Cases in Emotional and Behavioral Disorders in Children and Youth*. Pearson Education, Inc., Upper Saddle River, NJ, 2005.

The Universally Designed Classroom: Accessible Curriculum and Digital Technologies. 2005. Edited by Rose, David H., Meyer, Anne, and Hitchcock, Chuck. Harvard Education Press, Cambridge, MA 02138.

Vail, Priscilla L. Smart Kids with School Problems: Things to Know and How to Help. 1987. Penguin Books, New York, NY.

Future Books by Belinda Adams

Can You See Me Yet?

After completing Can You See Me Yet? Published in 2018, many people informed Belinda she had SO many more stories to tell. There were students this teacher remembered. There were hilarious antics left to share. There were unpredictable and sad situations that required the students be given a voice. If you haven't read the first book, no worries, because you can join Belinda on her journey of improving her craft of educating every type of student at any point. Bridging the gap between poverty and trauma, Belinda uses strategies she has learned through failures and success with her students and shares them with you.

Throw Away Kids

When most see those three words, it is the low-income and foster care situations that probably come to mind first. That is NOT what this book is about. Special education colleagues and diligent advocates for students with special needs, Carol Pirog and Belinda Adams, collaborate as they jump onto the slippery slope of issues plaguing students labeled as needing "special education".

Breaking protocols of what might be considered "politically correct", Carol and Belinda shed light on who these *Throw Away Kids* are and our obligation, as educators and parents, to turn the tide in their favor. *Throw Away Kids* are kids of every ethnicity and income range. They are the kids that need advocates to help them find their own voices and empower them as they move toward success.

About the Author

Belinda Adams is an elementary teacher. She has a bachelor's degree in elementary education and a master's degree in special education. She has taught students in kindergarten through eighth grade. Belinda has taught general education students, students in special education, and students in regular education who needed remediation.

Belinda is the author of *Can You See Me? Using Understanding to Help Students of Poverty Feel Seen, Heard and Valued in the Classroom* and *Don't Look Too Closely: What Children of Trauma are Hiding and How You Can Be the Difference.* Her books are based on her experiences in the classroom. With success rates that exceed expected yearly growth, Belinda is always available to discuss solutions for those dealing with difficult students because she believes every student *wants* to learn when provided the right motivation and support. Before becoming a teacher, Belinda earned a degree in psychology, worked in the business world, and wrote a weekly editorial for a northwestern Illinois newspaper.

In her spare time, oh wait, Belinda is a teacher and an author, she doesn't have spare time. She lives in the Midwest with her husband, her son, and her dogs, Murphy & Suzy Q.

Made in the USA
Monee, IL
13 September 2026

62981857R00059